YOUR KNOWLEDGE HAS VALUE

- We will publish your bachelor's and master's thesis, essays and papers

- Your own eBook and book - sold worldwide in all relevant shops

- Earn money with each sale

Upload your text at www.GRIN.com and publish for free

John Tarilanyo Afa

Soil Corrosion and Earthing

GRIN Verlag

Bibliografische Information der Deutschen Nationalbibliothek:

Die Deutsche Bibliothek verzeichnet diese Publikation in der Deutschen National-
bibliografie; detaillierte bibliografische Daten sind im Internet über http://dnb.d-
nb.de/ abrufbar.

Dieses Werk sowie alle darin enthaltenen einzelnen Beiträge und Abbildungen
sind urheberrechtlich geschützt. Jede Verwertung, die nicht ausdrücklich vom
Urheberrechtsschutz zugelassen ist, bedarf der vorherigen Zustimmung des Verla-
ges. Das gilt insbesondere für Vervielfältigungen, Bearbeitungen, Übersetzungen,
Mikroverfilmungen, Auswertungen durch Datenbanken und für die Einspeicherung
und Verarbeitung in elektronische Systeme. Alle Rechte, auch die des auszugsweisen
Nachdrucks, der fotomechanischen Wiedergabe (einschließlich Mikrokopie) sowie
der Auswertung durch Datenbanken oder ähnliche Einrichtungen, vorbehalten.

Imprint:

Copyright © 2011 GRIN Verlag GmbH
Druck und Bindung: Books on Demand GmbH, Norderstedt Germany
ISBN: 978-3-656-41265-6

This book at GRIN:

http://www.grin.com/en/e-book/213054/soil-corrosion-and-earthing

GRIN - Your knowledge has value

Der GRIN Verlag publiziert seit 1998 wissenschaftliche Arbeiten von Studenten, Hochschullehrern und anderen Akademikern als eBook und gedrucktes Buch. Die Verlagswebsite www.grin.com ist die ideale Plattform zur Veröffentlichung von Hausarbeiten, Abschlussarbeiten, wissenschaftlichen Aufsätzen, Dissertationen und Fachbüchern.

Visit us on the internet:

http://www.grin.com/

http://www.facebook.com/grincom

http://www.twitter.com/grin_com

JOHN TARILANYO AFA

CURRICULUM DESIGN

TECHNICAL REPORT

SCE 672:

SOIL CORROSION AND EARTHING

ATLANTIC INTERNATIONAL UNIVERSITY

HONOLULU, HAWAII

2011

Atlantic International University

TABLE OF CONTENT

SOIL CORROSION AND EARTHING

1.1   INTRODUCTION

The purpose of earthing electrical equipment or an installation is to ensure safety. No matter the sensitivity of a protective device, without proper earthing good performance of the device cannot be ensured. One of the major problems of having a continued low earthing resistance is corrosion. Corrosion is not limited to earth electrode only but all electrical equipments, buried metals, undergrounds telecommunication system, etc.

The life expectancy of a material or equipment depends on the severity of corrosion in the area (Gonos and Stathopuslos, 2006, Liu et al, 2004). The processes of corrosion are not simple and their effects are not easily predicted. Therefore if the corrosion processes are not controlled it will result to frequent and costly replacement of materials and equipments.

Due to the geographic location, the closeness to the ocean, and the soil conditions, the soil of Niger Delta are corrosive. This is particular true with the coastal soil. For this reasons, it was necessary to carry out a study of various soil structure and their corrosivity. Such records when properly presented will serve as a guide in the selection of earthing materials as well as help in the control of corrosion on buried materials.

1.1 Corrosion in Soil

All materials both organic and inorganic can react with their environments (Ala and Di Silvestre, 2002, Gonos and stathopuslos, 2006) and may eventually lose their usefulness for a given application.

Corrosion is the result of electrochemical, chemical or biological reaction between a metal and its surroundings. Corrosion of metal in soil is primarily electrochemical in nature and results from the operation of numerous galvanic corrosion cells (Sekioka et al, 2006, Laver and Griffiths, 2001, Liu et al, 2004).

Each galvanic corrosion cell comprises the following:

1.   An anode and cathode areas on metal surfaces

2.   Soil electrolyte

3.   Conducting path between an anode and cathode.

The quantity of metal lost by the corrosion process is directly proportional to the amount of direct current which flows through the corrosion cell. The weight loss was shown by Michael faraday as W=K*I*t

W=   Weight loss in grams

K=   Electrochemical equivalent in grams/ coulomb

I=   Current in Amperes

t=   Time in seconds.

For a given amount of current over a given period of time, the electrochemical equivalent (K) is the variable which determines the actual weight loss of the metal or material. Each metal has its own electrochemical equivalent which is a natural characteristic of that metal.

The thermodynamic instability of a metal depends on its amount of energy expended in extracting it from its natural ore (Ala et al, 2009, Saumade and Fontaine, 2008). The higher the energy expended, the higher is the thermodynamic instability, and the greater is the tendency of metal to migrate into electrotyle in ionic form. The thermodynamic instability of metals is expressed in terms of electrochemical potentials as shown in table1.

Table 1: Electrochemical Potential Series

Metal	Ion	Volts
Aluminium	Al^{+++}	-1.660
Zinc	Zn^{++}	-0.763

Iron	Fe^{++}	-0.440
Tin	Sn^{++}	-0.136
Lead	Pb^{++}	-0.126
Hydrogen	H^{+}	±0.000
Copper	Cu^{++}	+0.337

The magnitude of initial potential difference is influence by the electrochemical potential of metal and factors responsible for the formation of anodic areas. The greater the potential difference the higher the magnitude of corrosion current (Laver and Griffiths, 2001, Habjanic and Trlep, 2006).

$$I_o = \frac{E_{co} - E_{ao}}{R}$$

E_{co} = initial potential at cathode

E_{ca} = initial potential at anode

R = Ohmic resistance of the galvanic cell.

Ohmic resistance of the galvanic cell depends upon the specific resistivity of the electrolyte, ratio of the areas of the anodic and cathode phases and geometric configuration.

For a given corrosion cell, the effect of ohmic resistance can be predicted on the basis of resistivity of soil and distance between anode and cathode (Ala and Di Silvestre, 2002, Poljak and Doric, 2006).

There are two distinguishable corrosion cells, these are the micro and macro corrosion cells (Gupta, 2005, Mohamad et al, 2006).

Micro cell which causes corrosion of electrode (conductor) are formed due to heterogeneity in the metal surface and its composition. Physicochemical properties of soils are uniform at anode and cathode and the ohmic resistance in such cells is generally the same as predicted on the basis of physiochemical properties of soils.

Atlantic International University

The macro corrosion cells are mainly formed by the variation in the soil conditions along the surface of the metal. Difference in soil resistivity and differential aeration are major factors that are responsible for the formation of such cells.

Experimental data indicates that the degree of corrosion in most soil increases with decrease in resistivity of soil. If soil resistivity is low, the resistance path will be low and the magnitude of the galvanic current will be high. High galvanic current meant high rate of corrosion.

An increase in moisture content and salt content decreases the resistivity and increases corrosion. The generally accepted range of soil corrosivity as determined by electrical resistivity of soil and is given in table 2.

Table 2:   Soil Resistivity and Corrosion

Range of soil resistivity Ohm –meter	Class
Less than 25	Severely corrosive
25 – 50	Moderately corrosive
51 – 100	Mildly corrosive
Above 100	Very midly corrosive

Some investigators (Mohamad et al, 2006, Sekioka et al, 2006, gupta, 2005) considered that corrosivity of soil is directly related to its air penetration property and have postulated that poor aeration results in severe corrosion. This may be due to the fact that oxygen influences the static potential of metals. However, if there are differences in rate of air penetration through adjacent section of ground, well aerated areas will become the cathode and poorly aerated areas become the anode of galvanic corrosion cell.

The rate of air penetration and resistivity of soil decreases with increase in moisture content. In general, corrosivity of soil increases with moisture content up to a critical point, due to reduction

in ohmic resistance and beyond this point the corrosivity decreases due to reduction in oxygen supply required for cathode depolarization (Gupta, 2005). Moisture at the anode enhances the process there.

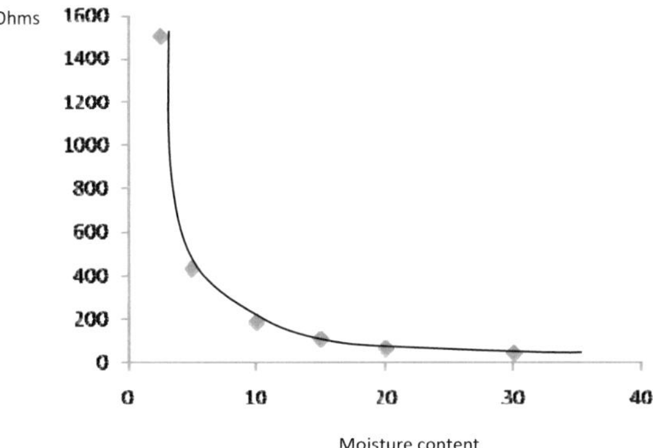

Fig. 1: Effect of water on earth resistance

In highly acidic soil, depolarization at the cathode increases due to availability of hydrogen and hence the corrosivity of the soil increases with acidity. The presence of salt reduces the resistivity of the soil and also may increase its heterogeneity. Acid salts increase corrosion and alkaline salts act as corrosion inhibitors.

2 General analysis

In order to actualize the study certain practical tests were carried out in order to compliment and compare with known results.

Due to the land formation in the Niger Delta, the area was divided into three main divisions. These main areas were the coastal areas, the inland areas and the upland areas. These are shown in table 3.

COPYRIGHT ANALYSIS - UD25640HCI34
SCIENCE AND Principles of Physics

Table 3: selected sites for resistivity test in Niger Delta

S/n	Coastal soil	In land	Upland
1	Bonny	Nembe	Port Harcourt
2.	Brass	Degema	Ahoda
3.	Kula	Okrika	Elele
4.	Opobo	Warri	Uyo
5.	Akassa	Yenagoa	Ugeli

2.1   Material and Method:

The four point method (Wenner Technique) was used for the earth resistivity measurement. This method uses four equally spaced short electrodes (1.5m) at equal distance of 5 meters in a single line. The outer two terminals of the instrument C_1 and C_2 were connected to the current electrodes and the two inner electrodes terminals P_1 and P_2 were connected to the potential terminals as shown in figure1.

The ratio of this measured potential to the calculated current for a given spacing is known as the apparent resistivity. That is, the resistivity of the soil is proportional to V/I.

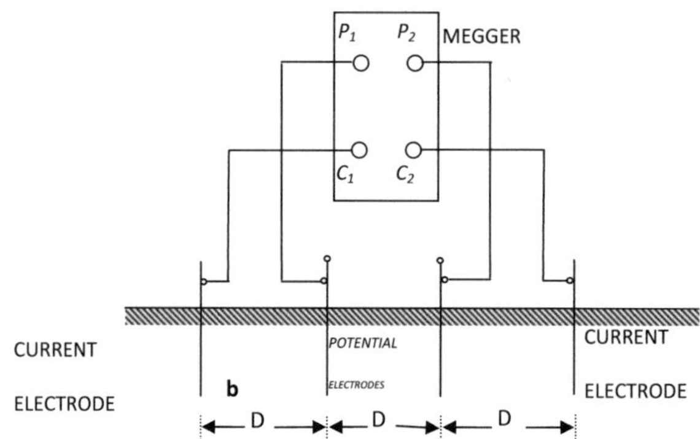

The resistance of soil can be calculated from eqn 2

$$\rho = 2\pi DR \qquad (2)$$

When b < < D

From a previous work done by the author the values of resistivities, average pH, and relative humidity was tested in twelve sites. The results are shown in table 4

Table 4: Resistivity Values of Soil, pH and Humidity

Site	Permanent moisture/Ground water level	Average Range of resistivity Ω-m	Average pH	Relative humidity (air) %
Bonny	1.0 meter	3.6 – 6	6.2	76%
Kula	0.8	3.2 – 4.0	6.1	82%
Brass	1.2	4.1 – 6.8	6.2	78%
Opobo	1.4	4.2 – 7.8	5.1	72%
Akassa	1.0	3.6 – 5.9	6.05	79%
Finima	1.3	3 – 6.2	6.0	77%
Degema	1.5	9.5 – 1.4	5.3	68%
Nembe	1.6	10 – 15.2	5.8	70%
Yenagoa	2.7	18 – 25	5.2	67%
Port Harcourt	3.5	22 – 32	4.3	63%
Ahoada	2.8	21 – 28	4.2	65%

Atlantic International University

Elele	3.1	22 – 28	4.4	62%

In other to update the values three more sites were tested as shown in fig.4. From the soil division three sites were selected, (Port Harcourt, Nembe and Brass) from each soil division and the resistivity values were tested at the depth of 0.8m, and 1.0m.

3 Results

The measurement results are presented in table 5.

Table 5: Resistivity Measurement

s/n	Test Sites	Resistivity in ohm-m		
		0.5m	0.8m	1.0m
1	Port Harcourt	410	230	160
2	Bonny	42	23	18
3	Nembe	92	42	25

From the measurement result graphs were presented as shown in fig2.

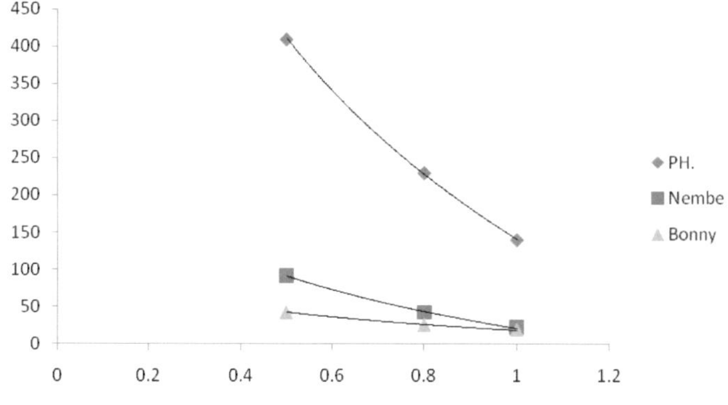

In other to confirm the permanent moisture levels two more reading were taken for the Port Harcourt site. There measurements were taken at depths of 2meters and 2.5 meters. From the results the permanent moisture level for Port Harcourt sites were placed between 2.5-3meters.

The sampled tests carried out on the sites are in close proximity with the previous tests, maintaining the authenticity of the previous result

4 Discussion:

In the coastal soil the resistivity values were between 10 Ω-m to 18 Ω-m. These are areas of high water table (1m-1.2m). The area has 60 percent of swamps and the rivers from the ocean overflow its bank. Due to this surrounding water, the water table at one meter depth is salty. There is therefore high corrosion rate in the soil.

The inland soil is the area where the coastal swamp soil terminates. Some settlements have the built-up areas extend to the swamp which is most times are the sand filled areas. Therefore the land mass is low.

The resistivity in this area is low especially at a depth of about 1.2 and below. The recorded value is about 15 Ω-m to 25 Ω-m, the values depends on the topography of the area and these values fall within the corrosive values given in the literatures.

The upland soil is sandy clay with permanent moisture level at about 3meters. The seasonal variation is more pronounced in this soil with high resistivity at the topsoil.

The resistivity values at the depth of 1.2meter is about 160 Ω-m but reduced drastically at 2.5meters. Judging from the level at 2.5meter this soil could be taken as corrosion. The pH values was about 4.4 -5.3 which could be described as acidic.

5 Recommendation:

Following the type of soil in the area and the various conditions of the place the following recommendations are made.

AIU ... EARH XXXX M.X ... UD2001SP02... 30
M.T of ... Soil Corrosion and Earthing

Atlantic International University

> Every electrode selected must have adequate Mechanical and Electrical properties to continue to meet the demands on them over a relatively long period of time. The material should have good electrical conductivity and should not easily corrode in the different soil.

> Coupling for copper rods need to have a minimum copper content of 80percent

> Use of dissimilar metals should be avoided but if used, the surfaces should be thoroughly cleaned and protected by an oxide inhibitor. Once the connection has been made the exterior should be protected by bitumastic paint or any other suitable material

> In planting electrodes like copper plated electrode, the one planted must be to the standard specification and in construction cracks must be checked or may result to bimetallic corrosion

> On no account must aluminum material be used either for earthing or connection to other electrode due to the risk of accelerates corrosion. The corrosion oxide layer is non conductive so could reduce the effectiveness of the earthing.

> Where copper electrode may form bimetal corrosion as a result of other buried materials in the soil, stainless steel rods are recommended.

> For the type of soil in the coastal soil where the soil may contain high chloride content the use Zinc coated electrode is not recommended.

6 Conclusion:

The coastal soil is extremely corrosive but from the readings of table 4 at a depth of 1.5-2.5 meters, all soil is within the range of very corrosion soil given in all available technical literatures. The salt content in the ground water at depth of 1.0meters, the marshy nature of the soil and the organic deposits contribute to the corrosion nature of the soil.

In the inland, soil most part of the soil ponds and form stagnant water at low tide. This encourages organic material depositions thereby having an acidic nature of the soil

The upland soil is also acidic but has a high resistivity value because of the water table (permanent moisture water table). From the measurement taken in September (rainy seasons)

the value of resistance was as low as 25 Ω-m at 1.5meters. This indicates that the effect of seasonal variation is pronounced.

It is very important to ensure that the earth electrode systems are correctly designed, installed and maintained.

From the soil characteristics the copper electrode is the recommended electrode in the area. This is due to it thermodynamic stability which makes it corrosion resistance in most corrosive soil and its low electrical resistance (good conductivity).

The copper plated electrode to the required standard could be used but due to careless handing and cracks on the electrode materials during installation, its used could be limited.

References

1. Afa, J.T., Anaele, C.M., 2010. Seasonal Variation of Soil Resistivity and Soil Temperature in Bayelsa State, American J. of Engrg. And Appl. Sci. 3 (4): 704 – 709.

2. Ala, G. and M.L. Di-Silvestre, 2002. A simulation model for electromagnetic transient in lightning protection systems on Eletromag. Comp. 44(4): 539 – 554.

3. Ala, G., M.L. Di-Silvestre, F. Viola and E. Francamano, 2009. Soil ionization due to high pulse transient current leaked by Earth electrodes Prog. Electromagn. Res. B., 14: 1- 21.

4. Ala, G., P. L. Buccheri, P. Romano, F. Viola 2008. Finite difference time domain simulation of earth electrodes soil ionization under lightning surge condition IET science measurement and technology 2(3): 134 – 145.

5. Gonos, I. F., A.X. Stathopuslos, 2006. Variation of soil resistivity and ground resistance during the year, 28th International conference on lightning protection. 740 – 744

6. Gupta, BR (2005) Power System Analysis and Design, S. Chand & Company LTD – New Delhi.

7. Laver, J. A. H., Griffiths, 2001. The variability of soil in Earthing Measurements and Earthing system performance; Rev. Energ. Ren: Power Engineering: 57-61.

8. Liu, Y. N., R. Theethayi, R. Thottappillu, R. M. Gonzales, M. Zitnik, 2004, An improved model for soil ionization around grounding system and its application to stratified soil; journal of electrostatics 60 (2- 4): 203 – 209.

9. Sekioka, S.M. I. Lorentzon, M. Philippakou, J. M. Prousalidis, 2006. current dependent grounding resistance model based on energy balance of soil ionization, IEEE Trans on power delivery 21(1) : 194 – 201.

10. Saumade, F; Fontaine, G (2008) Method and Apparatus for analysing the corrosive effect of the soil and its environment on a buried Metallic structure. United States Patent 4806850 2000.

11. Poljak, D. and V. Doric, 2006, "Wire antenna model for transient analysis of simple grounding systems, Part I: The vertical grounding electrode," *Progress In Electromagnetics Research*, Vol. 64, 149-166.

12. Habjanic, A. and M. Trlep, 2006. "The simulation of the soil ionization phenomenon around the grounding system by the finite element method," *IEEE Trans. on Magnetics* , Vol. 42, No. 4, 867-870.

13. Mohamad, N. N., A. Haddad, and H. Griffiths, 2006. "Characterization of ionization phenomena in soils under fast impulses," *IEEE Trans. on Power Delivery* , Vol. 21, No. 1, 353-361.

14. Zeng, R., X. Gong, J. He, B. Zang, and Y. Gao, 2008. "Lightning impulse performance of grounding grids for substations considering soil ionization," *IEEE Trans. on Power Delivery*, Vol. 23, No. 2, 667-675.

15. Sekioka, S., M. I. Lorentzou, M. P. Philippakou, and J. M. Prousalidis, 2006. "Current dependent grounding resistance model based on energy balance of soil ionization," *IEEE Trans. on Power Delivery*, Vol. 21, No. 1, 194-201.